Seid Yimer

Anaerobic Pond Design for Kombolcha Tannery

Seid Yimer

Anaerobic Pond Design for Kombolcha Tannery

Die Deutsche Bibliothek verzeichnet diese Publikation in der Deutschen Nationalbibliografie; detaillierte bibliografische Daten sind im Internet über http://dnb.d-nb.de/ abrufbar.

1. Auflage 2013
Copyright © 2013 GRIN Verlag GmbH
http://www.grin.com
Druck und Bindung: Books on Demand GmbH, Norderstedt Germany
ISBN 978-3-656-37579-1

WOLLO UNIVERSITY

INSTITUTE OF TECHNOLOGY

Chemical Engineering Department

Anaerobic Pond Design for Kombolcha Tannery

By
1. Seid Yimer Abate (M.Sc)
2. Temesgen Atnafu Yemata (M.Sc.)

February, 2013

WOLLO UNIVERSITY
INSTITUTE OF TECHNOLOGY
Chemical Engineering Department

Anaerobic Pond Design for Kombolcha Tannery

By

1. Seid Yimer Abate (M.Sc)
2. Temesgen Atnafu Yemata (M.Sc.)

A research submitted to the department of Chemical Engineering, Kombolcha Institute of Technology, Wollo University

Acknowledgement

First and foremost, praise to the Merciful Almighty God who blessed and taking care of us and our family and who enabled us accomplish this research work.

Next, we would like to express my deepest gratitude and sincere thanks to our chemical engineering department staff members and chemical engineering department secretary.

February, 2013

Seid Yimer
Temesgen Atnafu

Table of Contents

Contents

Anaerobic Pond Design for Kombolcha Tannery

List of Acronyms

BOD- Biological oxygen demand

COD-Chemical oxygen demand

$C_nH_aO_bN_d$ - Represents organic matter

C_t -Through flow rate

C_U -Unit capacity

F_{oa} - Open area factor

F_S -Slotted area

K_1 - First order kinetic constant

K_T -First order rate constant

L_i -BOD load at the inlet

Le-BOD load at the outlet

Ni - Inlet amount of faecal coliforms per 100ml

Ne -Outlet amount of faecal coliforms per 100ml

Q – Flow rate

SAR -Sludge accumulation rate

t -Retention time

Va -Anaerobic pond volume, m^3

λ_V -Volumetric BOD loading

Θa -Retention time

Θ_f- Retention time of facultative pond

Anaerobic Pond Design for Kombolcha Tannery

Abstract

Waste is the most horrible thing to our environment especially a waste from tannery industries. Kombolcha tannery since it is part of these industries it is affecting the people and brings drastic chance to the environment. Since this is the fact those treatment techniques are necessary to the factory like the one mentioned. In this study an aerobic pond is designed for tannery waste. It is possible to make suitable to our environment by treating it as per the type of the waste and degree of toxicity and hazard. This will be practiced only when the industrialists put themselves in to the boundaries of the industrial legislation otherwise good things will not happen unless the bad situation are developed. For the treatment techniques the one which has lower installation cost and lower running cost were used.

Key Words: *Anaerobic pond, Tannery waste, Hazard*

1. Introduction

1.1 Background

Domestic wastewater and most industrial wastewaters contain a mixture of various waste compounds of which the organic matter is an important part. In case of discharge of untreated wastewater into the environment, the organic matter is used in bacterial metabolism. This is called waste degradation or stabilization. In case of discharge of untreated wastewater into the environment, the organic matter is used in bacterial metabolism. This is called waste degradation or stabilization. Stabilization refers to the fact that after the organic matter has been degraded by bacteria, the bacterial activity based on the rest product is low. The material has become stable. Waste stabilization ponds are typically man-made basins surrounded by an earthen embankment. The waste is confined to the basin for a certain time span, such that bacteria that are present in the waste or added to it can stabilize the waste by degrading the organic fraction (Metcalf et al., 2003).

The advantages of waste stabilization ponds result from their simplicity. Ponds require relatively low capital investment when flat land is available at reasonable price. Ponds are easily maintained, can absorb shock loads, produce stabilized sludge that can periodically be removed and used in agriculture, whereas mechanical equipment to aerate or mix the sewage is hardly required. The consequent low operation and maintenance requirements, plus their potential to be easily scaled down to small scale application make them particularly attractive for remote towns and villages. All these features plus the ability to markedly reduce BOD, nutrients and pathogen concentrations have made waste stabilization ponds a very attractive treatment method (Jorge Jaramillo, 2003).

The most appropriate wastewater treatment is that which will produce an effluent meeting the recommended microbiological and chemical quality guidelines both at low cost and with minimal operational and maintenance requirements, Adopting as low a level of treatment as possible is especially desirable in developing countries, not only from the point of view of cost but also in acknowledgement of the difficulty of operating complex systems reliably. In many locations it will be better to design the reuse system to accept a low-grade of effluent rather than to rely on advanced treatment processes producing a reclaimed effluent which continuously meets a stringent quality standard (Metcalf et al., 2003).

Waste Stabilization Ponds are now regarded as the method of first choice for the treatment of wastewater in many parts of the world. Waste stabilization ponds are mainly shallow man-made basins comprising a single or several series of anaerobic, facultative or maturation ponds. The primary treatment takes place in the anaerobic pond, which is mainly designed for removing suspended solids, and some of the soluble element of organic matter (BOD_5). During the secondary stage in the facultative pond most of the remaining BOD_5 is removed through the coordinated activity of algae and heterotrophic bacteria. The main function of the tertiary treatment in the maturation pond is the removal of pathogens and nutrients (especially nitrogen). Waste stabilization pond technology is the most cost-effective wastewater treatment technology for the removal of pathogenic micro-organisms. The treatment is achieved through natural disinfection mechanisms. It is particularly well suited for tropical and subtropical countries because the intensity of the sunlight and temperature are key factors for the efficiency of the removal processes (Jeffrey P.J. et al., 1997).

1.2 Objectives

1.2.1 General Objectives

The general objective of this project is to design anaerobic pond for Kombolcha tannery.

1.2.2 Specific Objectives

- ❖ To characterize and determine the wastewater flow from Kombolcha tannery.
- ❖ To design waste stabilization pond
- ❖ To intend detail design of anaerobic pond

2. Waste in Leather Industry

Waste from leather industry consists of different kind of chemicals, tanned and untanned solid wastes (Prof. Dr. E. Heidemann, 1993).

2.1 Solid Waste

This includes cutting of untanned hides with hair, cutting of limed, shaved hair, flashing and cuttings. Some of the untanned solid wastes used for glue production factories for further processing and sometimes it is dumped on special dumps (K. Bienkiewicz et al., 1983).

2.2 Effluents

The composition of the effluent from leather factories is very complex and varies depending on the manufacturing procedure using in the factory concerned and all the starting materials .The particular characteristics of effluents from leather factories which start with row skin or hide are sulfide from the liming, high alkalinity, a very high proportion of dissolved organic compounds and in case of chrome leather factories chromium compounds. The effluent also contains certain amount of organic and inorganic suspended solids which from as a result of natural precipitation when the effluent from various processing stages are mixed. Official regulations that vary from country to country require that before any effluent is discharged to public water. It must be purified sufficiently to insure that there is no risk of biological equilibrium being disturbed where effluent is to be treated in public purification which might impair the functioning of this plants must first be removed (K. Bienkiewicz et al., 1983).

2.3 Harmful Effects of Pollutant Present in Tannery Waste

2.3.1 Chromium

- ➢ Corrosion effect in the initial tract
- ➢ Toxic dose for man
- ➢ Destructive of fish i.e. lethal limit to a stickle
  - Toxic in 30min, for gold fish
  - Toxic value for brume trout
- ➢ Impairment of photosynthesis of algae
- ➢ Distraction of microorganisms

- Toxic threshold effect in protozoa (Gerberei-Ing. Gerhard John, 1997).

2.3.2 Sulfide

- It is a most difficult martial in that it is highly toxic and has an obnoxious odor.
- Cause corrosion and may also release poisonous gas in the low parts of the sewer.
- Fatal to sensitive fish even in alkaline water (Gerberei-Ing. Gerhard John, 1997).

2.3.3 Organic

- Imposition high oxygen content
- Anthrax (T. C. Thorstensen,1976)

2.3.4 Suspended Mater

- Detraction fish food bottom found and spawning grounds of fish
- Choking sewer (T. C. Thorstensen, 1976).

2.3.5 PH

- Fishes are either killed or seriously handicapped
- Distraction to concrete structure (T. C. Thorstensen, 1976).

3. Tanning of Leather

Leather is a seldom used in its natural state as it's effected by variations in temperature (hard and stiff at low temperature, soft and limp at high temperature) liable to set. The purpose of tanning is to eliminate these problems with special vegetable, animal and synthetic substances. In preparing leather the process involved are used to remove the skin. In the process of removing undesired constituents of the skin the process used are named as bellow. These are curing, soaking, liming, fleshing, and deliming, bating, pickling, tanning. All these processes are done in order to get wet blue which is called semi-processed leather (Pocket Book for the Leather Technologist, 1997).

Tanning process is a slow process and in modern leather production involves a series of chemically inter dependent steps. Tanning is not only preserves the hide or skin but also make the leather resistance to cracking from flexing and the densile properties of leather depends on use (Pocket Book for the Leather Technologist, 1997).

3.1 Origin and Properties of Tanning Material

3.1.1Trimming

This is a process used to remove undesired part of the skin. This is part of solid waste in the tannery (J. H. Sharphouse, 1972).

3.1.2Washing and Soaking

Hide or skins when received by the tanning are usually in a condition of preservation based on dehydration. These skins may dried, as in the case with most goat skin and same tropical cattle hides or they may dehydrated by means of salt (J. H. Sharphouse, 1972).

Hair, dirty insecticide, blood and non-fibrous protein of the skin are removed and the moisture lost during preservation and storage is restored. Consequently the effluents are voluminous consists of a high percentage of solid, BOD and volatile solids with the PH usually being 7-8 (J. H. Sharphouse, 1972).

3.1.3 Liming

This process is accomplished by the use of lime with sodium sulfide thus effluents contain hair, high percentage of sulfide and high PH. This process is used to loosen the hair, remove the epidermis, and emulsify the skin grease and to bring about a more or less marked plumbing of the fibber structure (T. C. Thorstensen, 1976).

3.1.4 Fleshing

The lined skin is fleshed mechanically or by hand to remove epidemic resides the effluents from this process contain a high percentage of solids and to other processes water (Prof. Dr. E. Heidemann, 1993).

3.1.5 Deliming and Bating

After the operation deliming and bating are done to reduce the swelling, peptize the fiber and remove the protein degradation product thus making the skin reading for tanning as ammonium salts are used for bating and protein degradation. These two steps are further steps in the purification of the hide collagen network. Deliming is the removal of alkali and adjustment of the PH for bating. Bating is the enzymatic actions for the removal of unwanted hide complaints product are squeezed effluents of this process contain high volatile BOD and volatile solids(Prof. Dr. E. Heidemann, 1993).

3.1.6 Pickling

This process is necessary for chrome tanning to prevent the chances of chrome salt precipitation during tanning as the bates skins are treated with salt of this process that contain very high solid concentration and low PH of 3.0 (T. C. Thorstensen, 1976).

3.1.7 Chrome Tanning

During this process only 68.74 % for is utilized by the skins consequently a large amount of chrome surface rainout to the effluents color green while the PH lies in the range 2.5-2.8 (T. C. Thorstensen, 1976).

4. Waste Stabilization Ponds

4.1 Application of Waste Stabilization Pond Systems

Waste Stabilization Ponds are large, shallow basins in which raw sewage is treated entirely by natural processes involving both algae and bacteria. They are used for sewage treatment in temperate and tropical climates, and represent one of the most cost-effective, reliable and easily-operated methods for treating domestic and industrial wastewater. Waste stabilization ponds are very effective in the removal of faecal coliform bacteria. Sunlight energy is the only requirement for its operation. Further, it requires minimum supervision for daily operation, by simply cleaning the outlets and inlet works. The temperature and duration of sunlight in tropical countries offer an excellent opportunity for high efficiency and satisfactory performance for this type of water-cleaning system. They are well-suited for low-income tropical countries where conventional wastewater treatment cannot be achieved due to the lack of a reliable energy source. Further, the advantage of these systems, in terms of removal of pathogens, is one of the most important reasons for its use (Young et al., 1991).

4.2 Types of waste stabilization pond

Waste stabilization pond can be classified in respect to the type(s) of biological activity occurring in a pond. Three types are distinguished: anaerobic, facultative and maturation ponds. Usually a waste stabilization pond system comprises a single series of the aforementioned three ponds types or several such series in parallel. In essence, anaerobic and facultative ponds are designed for BOD removal (Biological Oxidation Demand) and maturation ponds for pathogen removal, although some BOD removal occurs in maturation ponds and some pathogen removal in anaerobic and facultative ponds. In many instances only anaerobic and facultative ponds are required. In general, maturation ponds are required only when stronger wastewaters (BOD > 150 mg/l) are to be treated prior to surface water discharge and when the treated wastewater is to be used for unrestricted irrigation (irrigation for vegetable crops). Generally, in waste stabilization pond systems, effluent flows from the anaerobic pond to the facultative pond and finally, if necessary, to the maturation pond. However, for better results wastewater flowing into an anaerobic pond shall be preliminary treated in order to remove coarse solids and other large materials often found in raw wastewater. Preliminary treatment operations typically include coarse screening, grit removal and, in some cases, comminution of large objects (Williams et al., 1998).

4.2.1 Aerobic Ponds

An aerobic stabilization pond contains bacteria and algae in suspension; aerobic conditions (the presence of dissolved oxygen) prevail throughout its depth. There are two types of aerobic ponds: shallow ponds and aerated ponds (Stenstrom et al., 1983).

4.2.1.1 Shallow Ponds

Shallow oxidation ponds obtain their dissolved oxygen via two phenomena: oxygen transfer between air and water surface, and oxygen produced by photosynthetic algae. Although the efficiency of soluble biochemical oxygen demand removal can be as high as 95 percent, the pond effluent will contain a large amount of algae which will contribute to the measured total biochemical oxygen demand of the effluent. To achieve removal of both soluble and insoluble biochemical oxygen demand, the suspended algae and microorganisms have to be separated from the pond effluent (Satyanarayan et al., 1987).

4.2.1.2 Aerated Ponds

An aerated pond is similar to an oxidation pond except that it is deeper and mechanical aeration devices are used to transfer oxygen into the wastewater. The aeration devices also mix the wastewater and bacteria. The main advantage of aerated ponds is that they require less area than oxidation ponds. The disadvantage is that the mechanical aeration devices require maintenance and use energy. Aerated ponds can be further classified as either complete-mix or partial-mix systems. A complete-mix pond has enough mixing energy input to keep all of the bacterial solids in the pond in suspension. On the other hand, a partial-mix pond contains a lesser amount of horsepower which is sufficient only to provide the oxygen required to oxidize the biochemical oxygen demand entering the pond (Satyanarayan et al., 1987).

4.2.2 Facultative Pond

Facultative pond is a combination of both aerobic and anaerobic pond. That is, in its top zone it is aerobic whereas it is anaerobic at its lower zone. Most of the stabilization pond constructed fall into the category of Facultative pond. Facultative ponds are of two types-primary facultative ponds and the secondary Facultative ponds. The primary facultative pond is one that receives raw wastewater whereas the secondary Facultative pond is one that receives the settled wastewater from the first stage. These ponds reduce BOD that helps in the production of algae that will further

Anaerobic Pond Design for Kombolcha Tannery

generate the oxygen needed to remove soluble BOD_5. The population of algae in any optimally performing facultative pond rest on organic load and temperature (Rosenwinkel et al., 1999).

Three zones exist in facultative pond. They are the following:
(1) A surface zone where aerobic bacteria and algae exist in a symbiotic relationship;
(2) An anaerobic bottom zone in which accumulated solids are actively decomposed by anaerobic bacteria; and
(3) An intermediate zone that is partly aerobic and partly anaerobic in which the decomposition of organic wastes is carried out by facultative bacteria. Because of this, these ponds are often referred to as facultative ponds. In these ponds, the suspended solids in the wastewater are allowed to settle to the bottom. As a result, the presence of algae is not required. The maintenance of the aerobic zone serves to minimize odor problems because many of the liquid and gaseous anaerobic decomposition products, carried to the surface by mixing currents, are utilized by the aerobic organisms (Rosenwinkel et al., 1999).

The amount of oxygen present in the pond depends upon the following factors:
- Temperature
- Organic loading
- Sunlight (Rosenwinkel et al., 1999).

These ponds are of two types: primary facultative ponds receive raw wastewater, and secondary facultative ponds receive the settled wastewater from the first stage (usually the effluent from anaerobic ponds). Facultative ponds are designed for BOD_5 removal on the basis of a low organic surface load to permit the development of an active algal population. This way, algae generate the oxygen needed to remove soluble BOD_5. Healthy algae populations give water a dark green color but occasionally they can turn red or pink due to the presence of purple sulphide-oxidizing photosynthetic activity. This ecological change occurs due to a slight overload. Thus, the change of coloring in facultative ponds is a qualitative indicator of an optimally performing removal process. The concentration of algae in an optimally performing facultative pond depends on organic load and temperature, but is usually in the range 500 to 2000 µg chlorophyll per liter. The photosynthetic activity of the algae results in a diurnal variation in the concentration of dissolved oxygen and pH values. Variables such as wind velocity have an important effect on the behavior of facultative ponds, as they generate the mixing of the pond liquid a good degree of mixing ensures a uniform distribution of BOD_5, dissolved oxygen, bacteria and algae, and hence better wastewater stabilization (Rajeshwari et al., 2000).

4.2.3 Maturation ponds

Maturation ponds are the ponds that receive effluent from a facultative pond. They are shallow, with less vertical stratification and are well oxygenated. The algae population of the maturation pond is much more diverse than that of facultative ponds. This clearly shows that the diversity of algae increases from pond to pond along the series. It is the Algae that cause the removal of pathogens and faecal coliforms. Though these ponds reduce small amount of BOD_5 but they play significant role in phosphorus and nitrogen removal (Annachhatre et al., 1996).

The main removal mechanisms especially of pathogens and faecal coliforms are ruled by algal activity in synergy with photo-oxidation. On the other hand, maturation ponds only achieve a small removal of BOD_5, but their contribution to nitrogen and phosphorus removal is more significant. A total nitrogen removal of 80% in all waste stabilization pond systems corresponds to 95% ammonia removal. It should be emphasized that most ammonia and nitrogen is removed in maturation ponds. However, the total phosphorus removal in WSP systems is low, usually less than 50% (Delgenes et al., 2003.

5. Anaerobic Ponds

5.1 Description of Anaerobic Pond

Anaerobic lagoons are typically used for two major purposes:

1) Pretreatment of high strength industrial wastewaters.

2) Pretreatment of municipal wastewater to allow preliminary sedimentation of suspended solids as a pretreatment process (Fabien Monnet, 2003).

Anaerobic lagoons have been especially effective for pretreatment of high strength organic waste waters. Applications include industrial wastewaters and rural communities that have a significant organic load from industrial sources. Biochemical oxygen demand (BOD) removals up to 60 percent are possible. The effluent cannot be discharged due to the high level of anaerobic byproducts remaining (Fabien Monnet, 2003).

Anaerobic ponds are deep treatment ponds that exclude oxygen and encourage the growth of bacteria, which break down the effluent. It is in the anaerobic pond that the effluent begins breaking down in the absence of oxygen "anaerobically". The anaerobic pond acts like an uncovered septic tank. Anaerobic bacteria break down the organic matter in the effluent, releasing methane and carbon dioxide. Sludge is deposited on the bottom and a crust forms on the surface. Anaerobic ponds are commonly 2-5 m deep and receive such a high organic loading (usually > 100 g BOD/m^3 d equivalent to > 3000 kg/ha/d for a depth of 3 m). They contain an organic loading that is very high relative to the amount of oxygen entering the pond, which maintains anaerobic conditions to the pond surface. Anaerobic ponds don't contain algae, although occasionally a thin film of mainly Chlamydomonas can be seen at the surface. They work extremely well in warm climate (can attain 60-85% BOD removal) and have relatively short retention time (for BOD of up to 300 mg/l, one day is sufficient at temperature $> 20^{\circ}$C)(Gerard Keily, 1997).

Anaerobic ponds reduce pathogenic microorganisms by sludge formation and the release of ammonia into the air. As a complete process, the anaerobic pond serves to:

✓ Separate out solid from dissolved material as solids settle as bottom sludge.
✓ Dissolve further organic material.

✓ Break down biodegradable organic material.

✓ Store undigested material and non-degradable solids as bottom sludge.

✓ Allow partially treated effluent to pass out (Lettinga et al., 1995).

These fermentation processes and the activity of anaerobic oxidation throughout the pond remove about 70% of the BOD_5 of the effluent. This is a very cost-effective method of reducing BOD_5. Normally, a single anaerobic pond in each treatment train is sufficient if the strength of the influent wastewater is less than 1000 mg/l BOD_5. For high strength industrial wastes, up to three anaerobic ponds in series might be justifiable but the retention time in any of these ponds should not be less than 1 day and Designers have been in the past too afraid to incorporate anaerobic ponds in case they cause odor. Formation of odor is strongly dependent on the type of waste to be treated in the plant, notably its sulphate (SO_4) concentration and volumetric loading rate, respectively. SO_4 is reduced to hydrogen sulphide (H_2S) under anaerobic conditions. H_2S is the compound mainly responsible for obnoxious odors. Other components besides H_2S and originating from the anaerobic decomposition of carbohydrates and proteins may contribute to obnoxious odors, too (Lettinga et al., 1991).

The HRT for ponds treating municipal sewage is between 1-3 days. For industrial applications HRT may increase to 20 days. In cold climates anaerobic ponds mainly act as settling ponds, whereas higher sewage temperatures enhance the anaerobic degradation process (hydrolysis, acid genesis, acetogenesis and methanogenesis). At higher temperatures BOD is therefore more effectively removed, especially the BOD-dissolved (Lettinga et al., 1991).

5.2 Treatment Mechanisms

Anaerobic lagoons process involves two separate but interrelated phases: acid formation and methane production. During the acid phase, bacteria convert complex organic compounds (carbohydrates, fats, and proteins) to simple organic compounds, mainly short-chain volatile organic acids (acetic, prop ionic, and lactic acids). The anaerobic bacteria involved in this phase are called "acid formers," and are classified as non -methanogenic microorganisms. During this phase, little chemical oxygen demand (COD) or biological oxygen demand (BOD) reduction occurs, because the short-chain fatty acids, alcohols, etc., can be used by many microorganisms, and thereby exerts an oxygen demand. The methane-production phase involves an intermediate step. First, bacteria convert the short chain organic acids to acetate, hydrogen gas, and carbon dioxide. This intermediate process is referred to as acetogenesis. Subsequently, several species of strictly anaerobic bacteria (methanogenic microorganisms) called "methane formers" convert the acetate, hydrogen, and carbon dioxide into methane gas (CH_4) through one of two major pathways. This process is referred to as methanogenesis. During this phase, waste stabilization occurs, represented by the formation of methane gas (Lusk et al., 1999). The two major pathways of methane formation are:

The breakdown of acetic acid to form methane and carbon dioxide:

$$CH_3COOH\ CH_4 + CO_2$$

The reduction of carbon dioxide by hydrogen gas to form methane:

$$CO_2 + 4H_2\ CH_4 + 2\ H_2O$$

The overall anaerobic decomposition of organic matter can be expressed by the following equation.

$$C_nH_aO_bN_d + (n-a/4-b/2+3d/4)H_2O \rightarrow (n/2+a/8-b/4-3d/8)CH_4 + (n/2-a/8+b/4+3d/8)CO_2 + dNH_3$$

where $C_nH_aO_bN_d$ represents organic matter (Lettinga et al., 1991).

The rate of anaerobic processes depends highly on temperature; in particular the methanogenic bacteria accelerate their metabolic activity with temperature. Optimum temperatures for growth are within the mesophyllic (20-45^0C) or thermophilic (40-65^0C) range. Substantial methane formation in anaerobic ponds is noticed at sewage temperatures above $15^0 C$ (Lettinga et al., 1991).

When the system is working properly, the two phases of degradation occur simultaneously in dynamic equilibrium. That is, the volatile organic acids are converted to methane at the same rate that they are formed from the more complex organic molecules. The growth rate and metabolism of the methanogenic bacteria can be adversely affected by small fluctuations in pH substrate concentrations, and temperature, but the performance of acid-forming bacteria is more tolerant over a wide range of conditions. When the process is stressed by shock loads or temperature fluctuations, methane bacteria activity occurs more slowly than the acid formers and an imbalance occurs. Intermediate volatile organic acids accumulate and the pH drops. As a result, the methanogens are further inhibited and the process eventually fails without corrective action. For this reason, the methane formation phase is the rate-limiting step and must not be inhibited. For the design of an anaerobic lagoon to work, it must be based on the limiting characteristics of these microorganisms (Liu et al., 2003).

Anaerobic Pond Design for Kombolcha Tannery

Table 5.1 Ideal operating range for methane fermentation (Liu et al., 2003)

Parameter	Optimum	Extreme
Temperature ^{O}C	30-35	25-40
pH	6.6-7.0	6.2-8.0
Alkalinity (mg/l)		
$CaCO_3$	2,000-3,000	1,000-5,000
Volatile acids (mg/l)		
Acetic acid	50-500	2,000

Concentrations such as calcium should be kept to a minimum. Excessive concentrations of these inhibitors produce toxic effects. Depending on its form, ammonia can be toxic to the bacteria as well as affect its concentration. Concentration of free ammonia in excess of 1,540 mg/L will result in severe toxicity, but concentrations of ammonium ion must be greater than 3,000 mg/L to produce the same effect. Maintaining a pH of 7.2 or below will ensure that most ammonia will be in the form of ammonium ion, so higher concentrations can be tolerated with little effect. The next table provides guidelines for acceptable ranges of other inhibitory substances (Lusk et al.,1999).

Table 5.2 Concentration of inhibitory substances (Metcalf et al., 2003).

Substance	Moderately inhibitor (mg/l)	Strongly inhibitor (mg/l)
Sodium	3,500-5,500	8,000
Potassium	2,500-4,500	12,000
Calcium	2,500-4,500	8,000
Magnesium	1,000-1,500	3,000
Sulfides	100-200	>200

5.3 Advantages and Disadvantages of Anaerobic Ponds

5.3.1 Advantages

- More effective for rapid stabilization of strong organic wastes, making higher influent organic loading possible.
- Produce methane, which can be used to heat buildings, run engines, or generate electricity, but methane collection increases operational problems.
- Produce less biomass per unit of organic material processed. Less biomass produced
- Equates to savings in sludge handling and disposal costs do not require additional energy, because they are not aerated, heated, or mixed.
- Less expensive to construct and operate
- Ponds can be operated in series (Nichols, C. E., 2004).

5.3.2 Disadvantages

- Require a relatively large area of land.
- Produce undesirable odors unless provisions are made to oxidize the escaping gases
- Require a relatively long detention time
- Environmental conditions directly impacts on operation (Nichols, C. E., 2004).

6. General Material Balance

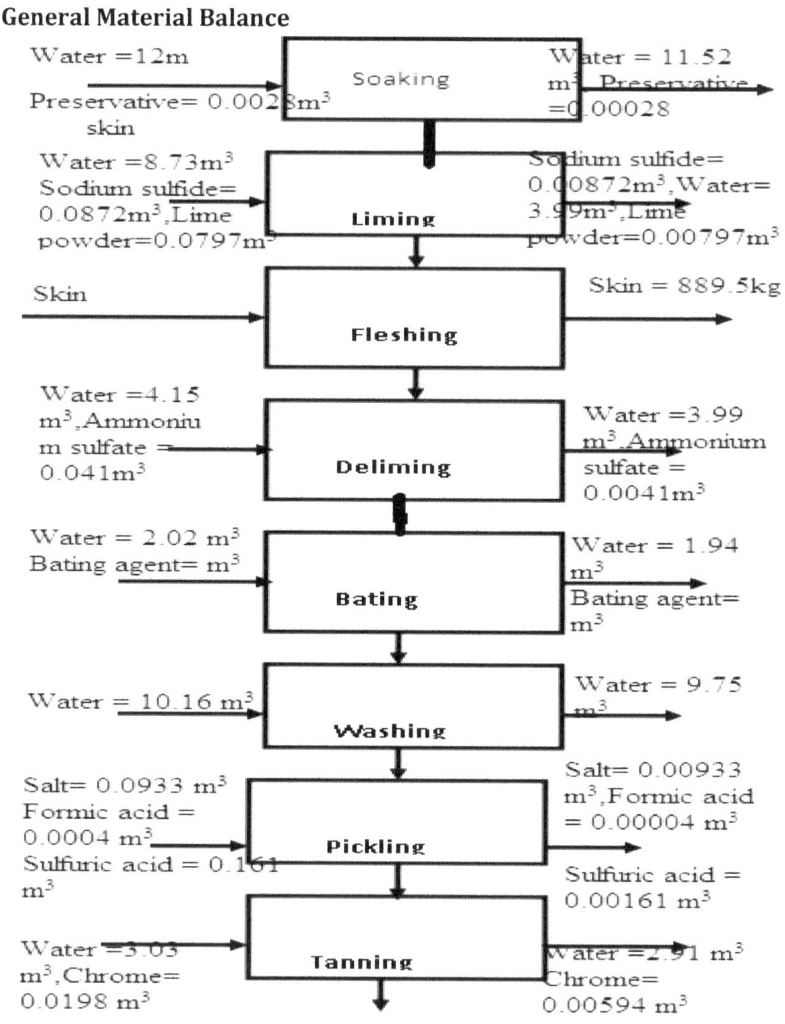

Fig 6.1 Material balance (Jorge Jaramillo, 2003)

6.1 Summary of the Material Balance

Table 6.1 Summary of material balance on soaking (Jorge Jaramillo, 2003)

Soaking	Input	Output
Water	12 m^3	11.52 m^3
Preservative	0.0028 m^3	0.00028 m^3

Table 6.2 Summary of material balance on unhairing(Jorge Jaramillo, 2003)

Unhairing	Input	Out put
Water	0.6 m^3	0.576m^3
Sodium sulfidrate	0.0298m^3	0.00298m^3
Sodium sulfide	0.0872m^3	0.00872m^3
Lime powder	0.0299m^3	0.00299m^3
Solid waste	-	90kg

Table 6.3 Summary of material balance on fleshing (Jorge Jaramillo, 2003)

Fleshing	Input	Output
Solid waste	-	889.5kg

Table 6.4 Summary of material balance on liming (Jorge Jaramillo, 2003)

Liming	Input	Output
Water	8.73m^3	8.38m^3
Lime powder	0.0797m^3	0.00797m^3

Table 6.5 Summary of material balance on deliming (Jorge Jaramillo, 2003)

Deliming	Input	Out put
Water	$4.15m^3$	$3.99m^3$
Ammonium sulfate	$0.041m^3$	$0.0041m^3$

Table 6.6 Summary of material balance on bating (Jorge Jaramillo, 2003)

Bating	Input	Output
Water	$2.02m^3$	$1.94m^3$

Table 6.7 Summary of material balance on washing(Jorge Jaramillo, 2003)

Washing	Input	Output
Water	$10.16m^3$	$9.75m^3$

Table 6.8 Summary of material balance on pickling (Jorge Jaramillo, 2003)

Pickling	Input	Output
Salt	$0.0933m^3$	$0.00933m^3$
Formic acid	$0.0004m^3$	$0.00004m^3$
Sulfuric acid	$0.0161m^3$	$0.00594m^3$

Table 6.9 Summary of material balance on tanning (Jorge Jaramillo, 2003)

Tanning	Input	Out put
Water	$3.03m^3$	$2.91m^3$
Chromium	$0.0198m^3$	$0.00594m^3$

Table 6.10 Summary of material balance on the major wastes from the tannery process(Jorge Jaramillo, 2003)

Anaerobic Pond Design for Kombolcha Tannery

Solid waste	979.8kg
Sulfide	$0.00872m^3$
Water	$39.066m^3$
Acid	$0.00165m^3$
Chromium	$0.00594m^3$
Salt	$0.01213m^3$

Total sum= $39.044m^3$

7. Pond Design

7.1 Flow Chart for the Treatment of Tanning Waste Water

The flow chart below illustrated the method for the treatment wastewater.

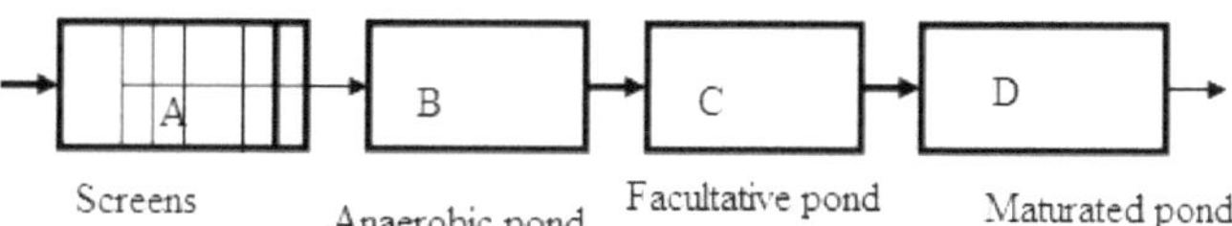

Fig 7.1 Flow chart for treating tanning waste water (Garcia-Heras, J.L., 2003).

1. Raw waste BOD =2300mg/lt

   SS =2800mg/lt

2. Treated effluent BOD <200mg/lt (Garcia-Heras, J.L., 2003).

7.2 Screen Design

Using through method of Mathew's equation

$$A = \frac{0.4 \times Ct}{Cu \times Foa \times Fs}$$

Where A   = screen area

C_t = through flow rate

C_U = unit capacity

F_{oa} = open area factor

F_S = slotted area (Garcia-Heras, J.L., 2003).

Let the screen is square opening specified by opening size

F_{oa} =   a/a+b = 5/6

= 0.69

Anaerobic Pond Design for Kombolcha Tannery

where d = diameter of wire

a = clear opening (Garcia-Heras, J.L., 2003)

* Assuming the diameter is 1 cm and the clear opening is 6cm

Through flow rate (the discharge rate of the effluent from the factory)

C_t = 40 m^3/d=4.63*10^{-4}m^3/s

Unit capacity (this can be read from data perry)

C_u = (using length to width ratio)

L/W=6/6=1(since it is a square openings)

C_U=2.2*10^{-5} (from the data reading) (Perry et al., 1997)

F_S is length to width ratio

F_S = 6/6 = 1

$$A = \frac{0.4 \times 4.63 \times 10^{-4}}{2.2 \times 10^{-5} \times 0.69 \times 1}$$

= 12.2 m^2 (Perry et al., 1997)

7.3 Anaerobic Pond Design

On the basis of volumetric BOD loading (λ_V, g/m^{3}d), which is given by:

Table 7.1 BOD loading (Poulsen et al., 2003)

Temperature (^{o}c)	volumetric loading (g/m^{3}d)	BOD removal (%)
<10	100	40
10-20	20T-100	2T+20
20-25	10T+100	2T+20
>25	350	70

Anaerobic Pond Design for Kombolcha Tannery

$$\lambda_V = L_iQ/V_a$$

where L_i = influent BOD, mg/l (= g/m^3)

$$Q = \text{flow, m}^3/\text{d}$$

$$V_a = \text{anaerobic pond volume, m}^3$$

Volumetric BOD loading can be calculated for a given temperature of 23^0 c

$$\lambda_V = 330 \text{ g/m}^3\text{d}$$

And the above equation can be re-arranged as

$$V_a = L_iQ /\lambda_V$$

$$V_a = \frac{2300\text{mg/l} \times 40\text{m}^3/\text{d}}{330\text{g/m}^3\text{d}}$$
$$= 280\text{m}^3 (\text{Poulsen et al., 2003}).$$

Pond area can be calculated by assuming the depth (D) =4m

$$A = V_a/D$$

$$A = 280\text{m}^3/4$$

$$A = 70\text{m}^2$$

Pond dimension can be calculated by considering length to width

Length: width   = 2:1

$$A = L \times W$$

$$L = 2W$$

$$A = 2W^2$$

$$W = \sqrt{70\text{m}^2/2}$$

$$= 5.916\text{m} \approx 6\text{m (Poulsen et al., 2003)}$$

Anaerobic Pond Design for Kombolcha Tannery

Retention time can be calculated

$$\Theta a = A/Q$$

$$= 280m^2/40m^3$$

$$\Theta a = 7 \text{ day (Poulsen et al., 2003)}$$

And the BOD removal can be calculated from the above table

$$\text{BOD removal} = 2 \times T + 20$$

$$\text{BOD removal} = 2 \times 23 + 20$$

$$\text{BOD removal} = 66mg/l \text{ (Poulsen et al., 2003)}$$

The performance of anaerobic ponds may deteriorate when ponds are getting full with sludge. The accumulated sludge causes the HRT to decrease and this may prevent complete settling and digestion of particulate matter. The BOD removal percentages calculated are valid only if the desludging is carried out in time that is when the pond volume is for one third full of sludge. The operational period of an anaerobic pond until desludging is required is given by

$$n = \frac{Va}{3} \times \frac{1}{Q \times SAR}$$

$$= \frac{280m^3}{3} \times \frac{1}{40m^3 \times 1.2}$$
$$= 1.944 \text{ year}$$

Where n = Operational period between desludging (years)

Q = inlet flow rate

SAR = Sludge accumulation rate, typically 1.2 m^3/year

V_{an} = Pond volume (Poulsen et al., 2003)

7.4 Facultative Pond

The facultative pond area can be calculated by kinetic model for designing

$$dL / dt = -K_1 L$$

$$L_e / L_i = 1 / (1 + K_1 t)$$

By rearranging the above equation

$$t = (L_i / L_e - 1)(1/K_1)$$

$$= (782/230 - 1)(1/0.3675)$$

$$= 8 \text{ day}$$

where, L_i = BOD load at the inlet

L_e = BOD load at the outlet

K_1 = first order kinetic constant

t = retention time (Mshandete et al., 2004)

Volume of the facultative pond can be calculated by

$$V_f = Q \times t$$

$$= 40 \times 8$$

$$= 320 \text{ m}^3 \text{ (Mshandete et al., 2004)}.$$

The area of the facultative pond can be calculated by taking the depth of 1.5m

$$A_f = V_f / D_f$$

$$= 320/1.5$$

$$= 213 \text{ m}^2 \text{ (Mshandete et al., 2004)}.$$

Using length to width ratio 2:1

$$A_f = 2 \times W^2$$

$W = 10.5$ m

$L = 21$ m

where, A_f =facultative pond area, m^2

V_f = volume of facultative pond

Q =flow rate in to the pond (Mshandete et al., 2004).

The cumulative filtered BOD removal in the anaerobic and facultative pond is 90% for T>20°C, so the facultative pond effluent has a filtered BOD of (0.1 x 2300), 230mg/l) (Mshandete et al., 2004).

7.5 Maturation Pond

For unrestricted irrigation assume the waste water contains (N_i= 5 x $10^{7)}$ faecal coliforms per 100 ml (Isaac A Doku, 2003).

K_T at the temperature of 23°C = 4.38 day^{-1} and the final outlet from the waste water (N_e = 1000) for unrestricted irrigation (Isaac A Doku, 2003).

Retention time for maturation pond can be calculated by:

$$\Theta m = \{[N_i/N_e \,(1 + K_T\,\Theta a)\,(1 + K_T\,\Theta_f)]^{\,1/n} - 1\}/\,K_T$$

$$= \{[5\, x10^7/1000\,(1 + 4.38\, x7)\,(1 + 4.38\, x\, 8)] - 1\}/\,4.38$$

$$= 10\ \text{day}$$

where K_T = first order rate constant

N_i = inlet amount of faecal coliforms per 100ml

N_e= outlet amount of faecal coliforms per 100ml

Θa= retention time of anaerobic pond

Θ_f= retention time of facultative pond

n = number of maturation pond(Isaac A Doku, 2003).

Anaerobic Pond Design for Kombolcha Tannery

By saying 90% of BOD loading is removed in anaerobic and facultative pond and the depth of 1m.

$$\lambda S = \frac{10(0.1 \times L_i) \times D}{\Theta m}$$

$$= \frac{10(0.1 \times 2300) \times 1}{10}$$

$$= 230 \text{ kg/ha day(Isaac A Doku, 2003)}$$

The effluent flow from the facultative pond can be calculated by

$$Q_i = Q - (0.001 \times A_f e)$$

$$= 40 - (0.001 \times 230 \times 4)$$

$$= 39.08 m^3/\text{day(Isaac A Doku, 2003)}$$

The area of maturation pond can be calculated

$$A_m = \frac{[2Q_i \Theta m]}{[2D + 0.001 e \Theta m]}$$

$$= \frac{[2 \times 39.08 \times 4]}{[2 \times 1 + 0.001 \times 4 \times 4]}$$

$$= 157.58 \text{ m}^2 \text{ (Isaac A Doku, 2003)}.$$

Using length to width ratio of 5:1

$$A_m = L \times W$$

$$= 5 \times W^2$$

$$W = 5.6 \text{ m}$$

$$L = 28 \text{ m(Isaac A Doku, 2003)}$$

The effluent flow from the maturation pond is

$$Q_{im} = Q_i - (0.001 \times A_m e)$$

$$= 39.08 - (0.001 \times 157.58 \times 4)$$

$$= 38.45 \ m^3/ \ day$$

Thus only 4% of flow is lost due to evaporation(Isaac A Doku, 2003).

The BOD removal of the cumulative of anaerobic and facultative ponds is 90% of redaction and in the maturation pond the BOD loading is reduced by 25%, so the final outlet from the maturated pond can be calculated as follows:

$$L_{if} = 0.1 \times 0.75 \times L_i$$

$$= 0.1 \times 0.75 \times 2300$$

$$= 172.5 \ mg/l(\text{Isaac A Doku, 2003}).$$

The effluent of the waste stabilization pond wall treated and it is below the minimum amount of BOD load to be discharged in the river (Isaac A Doku, 2003) .

7.6 Detail Design of Anaerobic Pond

The volume of the pond

$$Va = 280 \ m^3$$

The area of anaerobic pond by taking the pond depth

$$D = 4 \ m$$

$$A = 70 \ m^2$$

The length of the pond by taking length to width ratio of 2:1

$$L = 12 \ m$$

$$W = 6 \ m$$

Retention time of the pond

$$\Theta a = 7 \ day \ (\text{Malina, Jr. J.F., 1992}).$$

Material of constriction for anaerobic pond is concrete and the thickness of the concrete can be calculated by

$$T = 30 \times D + 50mm$$

$$= 30 \times 4 + 50$$

$$= 170 \text{ mm}$$

where,

T = Thickness (mm)

D = depth (m) (Malina, Jr. J.F., 1992).

Embankment slopes are commonly 1 to 3. Steeper slopes may be used if the soil is suitable; slope stability should be ascertained according to standard soil mechanics procedures for small earth dams. Embankments should be planted with grass to increase stability: a slow-growing rhizomatous species should be used to minimize maintenance. The minimum free board that should be provided is decided on the prevailing waves, induced by the wind, from overtopping the embankment. For small ponds the minimum free board is 0.5 m (Malina, Jr. J.F., 1992).

7.7 Inlet outlet structure of anaerobic pond

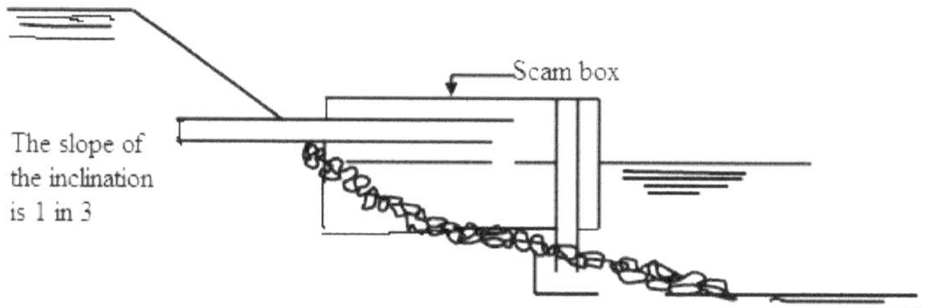

Fig 7.2   Inlet structure of anaerobic pond (Jorge Jaramillo, 2003),

8. Pond Layout

The basic units in waste stabilization pond systems are anaerobic, facultative and maturation ponds. These ponds can be put together in various ways to develop a successful treatment system. Anaerobic ponds are put in parallel to provide operational flexibility for sludge removal and improve the distribution of settled solids over the entire available pond surface. Facultative or maturation ponds are put in series to achieve a better overall performance with respect to BOD removal and faecal coliform reduction. Some degree of pre-treatment of raw sewage is desirable to remove large solids that would otherwise cause clogging of interconnecting pipe work or would cause undesired scum layer formation in the facultative and maturation ponds (Jeffrey P.J. et al., 1997).

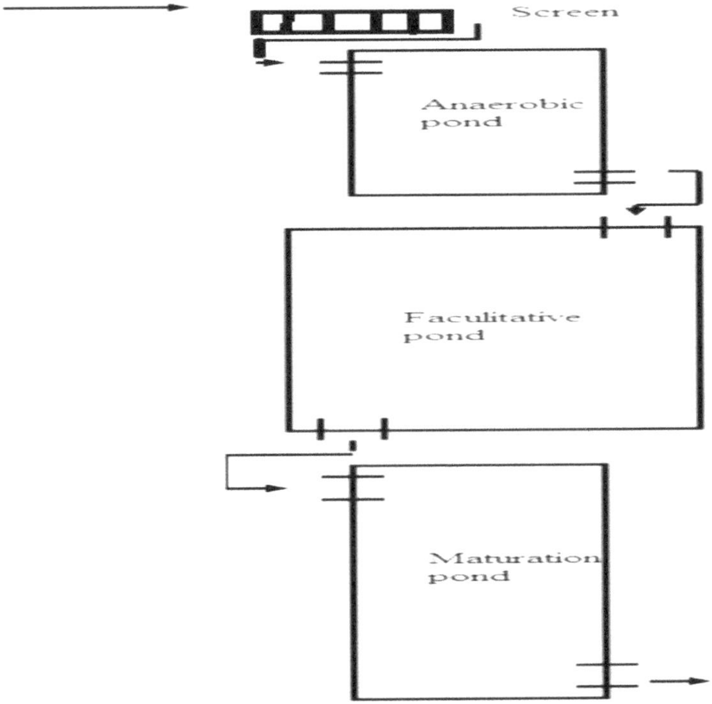

Fig 8.1 Layout on waste stabilization pond (Jeffrey P.J. et al., 1997).

8.1 Pond Geometry

There has been little rigorous work done on determining optimal pond shapes. The most common shape is rectangular, although there is much variation in the length-to-breadth ratio. Clearly, the optimal pond geometry, which includes not only the shape of the pond but also the relative positions of its inlet and outlet, is that which minimizes hydraulic short-circuiting. In general, anaerobic and primary facultative ponds should be rectangular, with length-to-breadth ratios of 2 - 3 to 1 so as to avoid sludge banks forming near the inlet. However, the geometry of secondary facultative and maturation ponds is less important than previously thought; they can have higher length-to-breadth ratios (up to 10 to 1) so that they better approximate plug flow conditions. Ponds do not need to be strictly rectangular, but may be gently curved if necessary or if desired for aesthetic reasons. A single inlet and outlet are usually sufficient, and these should be located in diagonally opposite corners of the pond (the inlet should *not* discharge centrally in the pond as this maximizes hydraulic short-circuiting). The use of complicated multi-inlet and multi-outlet designs is unnecessary and not recommended (Jeffrey P.J. et al., 1997).

To facilitate wind-induced mixing of the pond surface layers, the pond should be located so that its longest dimension (diagonal) lies in the direction of the prevailing wind. If this is seasonally variable, the wind direction in the hot season should be used as this is when thermal stratification is at its greatest. To minimize hydraulic short-circuiting, the inlet should be located such that the wastewater flows in the pond against the wind. Baffles should only be used with caution. In facultative ponds, when baffles are needed because the site geometry is such that it is not possible to locate the inlet and outlet in diagonally opposite corners, care must be taken in locating the baffle(s) to avoid too high a BOD loading in the inlet zone (and consequent possible risk of odour release). In maturation ponds baffling is advantageous as it helps to maintain the surface zone of high pH, which facilitates the removal of faecal bacteria. The areas calculated by the process design procedure described are mid-depth areas, and the dimensions calculated from them are thus mid depth dimensions. These need to be corrected for the slope of the embankment (Jeffrey P.J. et al., 1997).

A more precise method is advisable for anaerobic ponds, as these are relatively small. The following formula is used.

$$V_a = [(LW) + (L - 2s\ D)\ (W - 2s\ D) + 4\ (L - s\ D)\ (W - s\ D)]\ [D/6]$$

where,

V_a = anaerobic pond volume, m^3

L = pond length at TWL, m

W = pond width at TWL, m

s = horizontal slope factor (2.2)

D = pond liquid depth, m. With the substitution of L as nW, based on a length to breadth ratio of n to 1, becomes a simple quadratic in W (Malina, Jr. J.F., 1992).

The dimensions and levels that the contractor needs to know are those of the base and the top of the embankment; the latter includes the effect of the freeboard. The minimum freeboard that should be provided is decided on the basis of preventing waves, induced by the wind, from overtopping the embankment. For small ponds (under 1 ha in area) 0.5 m freeboard should be provided; for ponds between 1 ha and 3 ha, the freeboard should be 0.5-1 m, depending on site considerations. The depth chosen for any particular pond depends on site considerations (presence of shallow rock, minimization of earthworks). The depth of facultative and maturation ponds should not be less than 1m so as to avoid vegetation growing up from the pond base, with the consequent hazard of mosquito and snail breeding (Malina, Jr. J.F., 1992).

8.2 Inlet and Outlet Structure

There is a wide variety of designs for inlet and outlet structures, and provided they follow certain basic concepts, their precise design is relatively unimportant. Firstly, they should be simple and inexpensive; while this should be self-evident, it is all too common to see unnecessarily complex and expensive structures. Secondly, they should permit samples of the pond effluent to be taken with ease. The inlet to anaerobic and primary facultative ponds should discharge well below the liquid level so as to minimize short-circuiting (especially in deep anaerobic ponds) and thus reduce the quantity of scum (which is important in facultative ponds). Inlets to secondary facultative and maturation ponds should also discharge below the liquid level, preferably at mid-depth in order to reduce the possibility of short-circuiting (Mahony et al. 2002).

The outlet of all ponds should be protected against the discharge of scum by the provision of a scum guard. The take-off level for the effluent, which is controlled by the scum guard depth, is important as it has a significant influence on effluent quality. In facultative ponds, the scum guard should extend just below the maximum depth of the algal band when the pond is stratified so as to minimize the daily quantity of algae, and hence BOD, leaving the pond. In anaerobic and maturation ponds, where algal banding is irrelevant, the take-off should be nearer the surface: in anaerobic ponds it should be well above the maximum depth of sludge but below any surface crust, and in maturation ponds it should be at the level that gives the best possible microbiological quality. The following effluent take-off levels are recommended:

Anaerobic ponds: 300 mm

Facultative ponds: 600 mm

Maturation ponds: 50 mm (Malina, Jr. J.F., 1992).

8.3 Inlet Outlet Structure of Anaerobic Pond

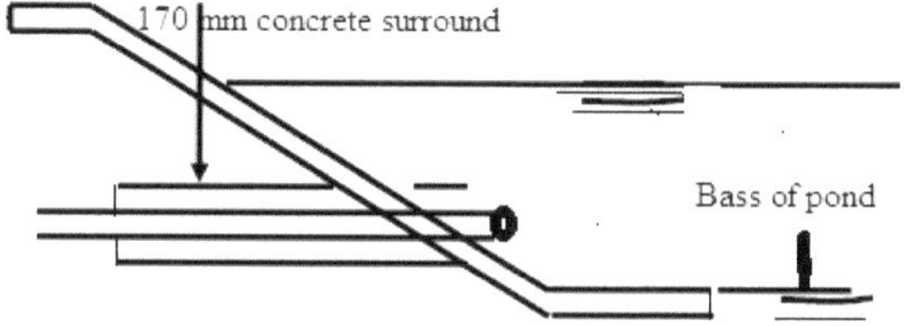

Fig 8.2 Inlet structure of anaerobic pond (Malina, Jr. J.F., 1992).

Fig 8.3 Inlet structure for facultative and maturation ponds (Malina, Jr. J.F., 1992).

9. Chrome Removal in Anaerobic Pond

The main pollutant in leather processing industry is chrome which affects the living things in discharging water body (Prof. Dr. E. Heidemann, 1993).

The wastewater resulting from these processes contains high amount of chromium metal which is harmful for environment and human health. Tanning process using chromium compounds is one of the most common methods for processing of hides. In this process about 60% -70% of chromium reacts with the hides. In the other word, about 30%- 40% of the chromium amount remains in the solid and liquid wastes (especially spent tanning solutions). Hence, the wastewater of tanning process is an important source adding Cr pollutant to the environment. In addition, the cost of the chromium metal is also important and it is possible to be recovered from the wastewater and reported that the Cr ions concentration in the tanning wastewater varies from 2500 to 8000 ppm and 1300 to 2500 ppm respectively. Several methods have been used for removing toxic metal ions from aqueous solutions. These include chemical precipitation, ion exchange, reverse osmosis, membrane processes, evaporation, solvent extraction, and adsorption of these; chemical precipitation is the usual way for this purpose (T. C. Thorstensen, 1976).

Many factors affect the process of chemical precipitation including the type of precipitation agent, pH, and velocity of precipitation, sludge volume, time of mixing and complexing agents. Using of chrome in the tanning process will choose:

> Chromium that enters the tannery waste streams represents the loss of a valuable raw material; and

> Chromium in the tannery waste streams also represents an increasingly expensive waste treatment problem (T. C. Thorstensen, 1976).

The best precipitant agent for removing chrome from anaerobic pond is lime (NaOH) which have vary low sludge volume, the settling rate, and also increase the PH of waste water which help to decrease the chrome and also prevent the formation of odorous (H_2S) (T. C. Thorstensen, 1976).

9.1 Site Selection

A land fill has to meet several locations and geo technical design criteria and be acceptable to the public for these purpose usually a circle indicating a search radius "(the maximum distance a waste generating willing to the waste) is drown on the road map of the region .keeping the waste generator (the industry) at it's center, the "search" radius should be depending on the economics of the waste howling. Howling of waste is one of the high cost inland fill operation. So it is essential to keep the cost as low as possible and should start it with small search radius and enlarge if it needed. Final site selection of disposal site is usually based on the result of a preliminary site survey i.e based on results of engineering design and cost studies, and environmental impact assessment (Hughes, D.E., 1980).

Anaerobic Pond Design for Kombolcha Tannery

Table 9.1 Factors that must be considered fin evaluating potential landfills sites (Hughes, D.E., 1980).

Factor	Remarks
Available land area	Site should have a useful life greater than 1 year (minimum value)
Haul distance cost operating	Will have a significant impact on
Soil condition and topography near the site	Cover material must be available oat or
Surface water hydrology	Impacts drainage requirements
Geological and hydrology conditions	Probably most important factors in establishment of land fill site especially with respect to site operations.
Climatologic conditions	Provisions must be made for wet weather conditions.
Local environmental conditions	Noise, odor, dust vector and aesthetic factors
Ultimate use of site	Control requirements affect long term management for site.

Anaerobic Pond Design for Kombolcha Tannery

Table 9.2 Safe distance used (Hughes, D.E., 1980).

Criteria distance	Source
300m away from	Lake or pond
90 m " "	River
300m " "	High way
300m " "	Public parks
3048m " "	Airports
365 m " "	Water supply well

10. Cost Analysis

The primary cost associated with constructing an anaerobic pond is the cost of the land, earthwork appurtenances, and required service facilities. Costs for forming the embankment, compacting, lining, service road and fencing, and piping and pumps also need to be considered. Operating costs and power requirements are minimal. The land cost of the waste stabilization pond of the area 441 m^2 is 200,000 birr (Matches Process Equipment Cost estimates, 2007).

Material of constriction for anaerobic pond is concrete and the thickness of the concrete can be calculated by

$$T = 30 \times D + 50mm$$

$$= 30 \times 4 + 50$$

$$= 170 \ mm$$

where, T = Thickness (mm)

D = depth (m) (Garcia-Heras, J.L., 2003).

The volume of the concrete used to construct the pond using the pond thickness calculated as follows

$$V_{concrete} = 4 \times T \times L \times D$$

$$V_{concrete} = 4 \times 0.17 \times 12 \times 4$$

Anaerobic Pond Design for Kombolcha Tannery

$V_{concrete}$ = 32.64 m³ (Garcia-Heras, J.L., 2003).

The total volume required to excavate will be the sum of this volume and the volume of the pond can be found by

$$V_{excavate} = V_{total} + 50$$

$$= 757.58 + 50$$

$$V_{excavate} = 807.58 \text{ m}^3 \text{ (Garcia-Heras, J.L., 2003)}$$

Table 10.1 Present estimated costs (Matches Process Equipment Cost estimates, 2007)

Type of work	Standard cost (birr per cubic meter)	Total estimated cost
Concrete	320.00	5222.4
Excavation	25.00	20189.5

The base of pond has the thickness of 150 mm and the volume of the concrete will be

$$V = L \times W \times T$$

$$V = 6 \times 12 \times .15$$

$$V = 10.8 \text{ m}^3$$

The cost of the concrete using the above table is equal to 3456 birr. From this the total cost needed to construct one pond is estimated as the sum of both this costs which is equal to 28867.9 birr (Matches Process Equipment Cost estimates, 2007).

Estimating the cost of facultative pond by the same method:

$$T_f = 2 \times D + 50$$

$$= 1.5 \times 30 + 50$$

$$= 95 \text{ mm(Sinnott, R.K., 1999)}$$

The thickness of the wall is too thin so we can take the minimum thickness of the wall which is equal to 150 mm.

Anaerobic Pond Design for Kombolcha Tannery

$$V_{concrete} = 2 \times T \times L \times D$$

$$= 9.45 \text{ m}^3$$

$$V_{excavate} = V_{total} + 50$$

$$= 59.45 \text{ m}^3$$

$$V_{baes} = L \times W \times T$$

$$= 33.075 \text{ m}^3 (\text{ Sinnott, R.K., 1999})$$

The total cost of facultative pond is the sum of three costs 15094.25 birr (Matches Process Equipment Cost estimates, 2007).

Estimating the cost maturation pond

$$T_m = 2 \times D + 50$$

$$= 55 \text{ mm}$$

The thickness of the wall is too thin so we can take the minimum thickness of the wall which is equal to 150 mm (Sinnott, R.K., 1999).

$$V_{concrete} = 4 \times T \times L \times D$$

$$= 94.08 \text{ m}^3$$

$$V_{excavate} = V_{total} + 50$$

$$= 97.04 \text{ m}^3$$

$$V_{baes} = L \times W \times T$$

$$= 23.52 \text{ m}^3 (\text{Sinnott, R.K., 1999}).$$

By using the above table to estimate the cost the total cost of maturation pond is equal to 25005.2 birr (Matches Process Equipment Cost estimates, 2007).

The total cost for buying the land and constructing the ponds for the treatment of tanning waste is equal to the sum of :

Cost of anaerobic pond = 28867.9 birr

Cost of facultative pond =15094.25 birr

Cost of maturation pond = 25005.2 birr

Cost of land for treatment= 200000 birr

Total cost –268967.35 birr (Matches Process Equipment Cost estimates, 2007).

11. Conclusion and Recommendation

11.1 Conclusion

Tannery industries waste is the worst thing to our environment. It is possible to make suitable to our environment by treating it as per the type of the waste and degree of toxicity and hazard. This will be practiced only when the industrialists put themselves in to the boundaries of the industrial legislation otherwise good things will not happen unless the bad situation are developed.

Kombolcha tannery since it is part of these industries it is affecting the people and brings drastic chance to the environment. Since this is the fact those treatment techniques are necessary to the factory like the one mentioned.

The treatment techniques we preferred is the one which has lower installation cost and lower running cost and they doesn't need a professional person to operate them. They only need good management to allocate budget as per the requirement of the design.

In addition to the treatment techniques glue manufacturing from fatty protein like fleshing, is one of income generating media which may support the treatment expense to some extent.

11.2 Recommendation

At this project we tried to point out some environment problems of Kombolcha tannery and we tried to point out some solutions for those problems. However we need to recommend some additional facts.

Anaerobic Pond Design for Kombolcha Tannery

- We would like to recommend clean technology options such as recycling of chrome from tanning process, which will save our environment and the factories expense as well.
- We tried to construct anaerobic pond without a gas collector but we recommend a gas collector to collect methane and to apply waste to energy principle providing we put the amount of methane produced.

Finally we need to recommend treating the trimmings with lime and producing glue like fatty fleshing.

References

Annachhatre, A.p.,1996, Anaerobic Treatment of Industrial Wastewater, Resources conservation, and recycling, vol, 16, pp.161-166.

Delgenes, J.P., Penaud, V. & Moletta, R.,2003, Pretreatments for the enhancement of anaerobic digestion of solid wastes. In: Biomethanization of the Organic Fraction of Municipal Solid Wastes (edited by J. Mata-Alvarez). Pp. 201-228. London: IWA

Fabien Monnet, 2003, An introduction to anaerobic digestion of organic wastes, final report, Scotland.

Garcia-Heras, J.L., 2003, Reactor sizing, process kinetics, and modelling of anaerobic digestion of complex wastes. In: Biomethanization of the Organic Fraction of Municipal Solid Wastes (edited by J. Mata-Alvarez). Pp. 31-43.

Gerard Keily, 1997, Environmental Engineering. Mc Grow Hill, London.

Gerberei-Ing. Gerhard John Auflage, 1997, Possible defects in leather production Selbstverlag -D- 68623 Lampertheim

Isaac A Doku, 2003, The potential for the use of upflow anaerobic sludge blanket reactor for the treatment of faecal sludge in Ghana.

Jeffrey P.J., Aarne, V. P., and Ruth, F.W., 1997, Environmental pollution and control, 4th ed., Elsevier science and technology publisher.

J. H. Sharphouse,1972, Leather Technician's Handbook 2nd edition Leather Producers' Association – London/UK

Jorge Jaramillo,2003, Guidelines for the Design, Construction and Operation of Manual Sanitary Landfills, Universidad de Antioquia, Colombia.

Anaerobic Pond Design for Kombolcha Tannery

K. Bienkiewicz, R. E. Krieger,1983,Physical Chemistry of Leather Making 1st edition Publishing Company-Malabar Florida/USA

Lettinga, G., 1995, Anaerobic digestion and wastewater treatment systems, Antonie van Leeuwenhoek, vol. 67, pp. 3-28.

Lettinga,g, and L.W. Hulshoff pol, 1991,Anaerobic Sludge Blanket Process Designs for various types of Wastewater" Water Science and Technology, vol24,no8..

Liu, Y., Xu, H.-L., Yang, S.-F. & Tay, J.-H., 2003, Mechanisms and models for anaerobic granulation in upflow anaerobic sludge blanket reactor. Water Research, 37(3), 661-673.

Lusk, P., 1999, Latest Progress in Anaerobic Digestion, Biocycle vol. 40 (7).

Matches' Process Equipment Cost estimates, 2007. http://www.matche.com/EquipCost/index.htlm, retrieved on April 29, 2009.

Mahony, T., and O'Flaherty, V. et al. 2002. Feasibility Study for Centralised Anaerobic digestion for the Treatment of Various Wastes and Wastewaters in Sensitive Catchment Areas, Ireland Environmental Protection Agency.

Malina, Jr. J.F., 1992. Anaerobic sludge digestion, In: design of anaerobic processes for treatment of industrial and municipal wastes, water quality management libraryvolume 7, technomic publishing company, Pennsylvania, USA.

Metcalf and Eddy, 2003,Waste water engineering treatment and reuse, 4th ed. Tata McGraw-Hill publisher: New Delhi.

Nichols, C. E., 2004, Overview of Anaerobic Digestion Technologies in Europe, Biocycle 45(1), pp. 47-54.

Perry, R. H., Green, D. W. and Maloney, J. O. (eds) ,1997, Perry's Chemical Engineers Handbook, 7th edn. (McGraw-Hill).

Poulsen, T.G., 2003, Anaerobic digestion, solid waste management, Aalborg University, Denmark.

Pocket Book for the Leather Technologist, 1997, fourth edition, revised and enlarged BASF Aktiengesellschaft 67056 Ludwigshafen Germany

Prof. Dr. E. Heidemann, 1993, Fundamentals of Leather Manufacturing 1st edition Eduard Roether KG – D-64212 Darmstadt

Rajeshwari, K.V., Balakrishnan, M., Kansal, A., Lata, K. & Kishore, V.V.N, 2000, State-of the-art of anaerobic digestion technology for industrial wastewater treatment. Renewable and Sustainable Energy Review, 4(2), 135-156.

Rosenwinkel, K.H., and Meyer, H., 1999, Anaerobic treatment of slaughterhouse residues in municipal digesters, water science and technology vol. 40 (1), pp. 101-111.

Satyanarayan, S., Kaul, S. N., Badrinath, S. D. and Gadkari, S. K. 1987. Anaerobic treatment of human wastes, Indian Journal of Environmental Protection, Vol. 7, No. 1, pp. 5-10.

Sinnott, R.K., 1999, Chemical Engineering Design, Coulson and Richardson's Chemical engineering, 3rd ed., volume 6, Butterworth-Heinemann Elsevier science publisher, Great Britain.

Stenstrom, M.K., Ng, A.S., Bhunia, P.K., and Abramson, S.D., 1983,Anaerobic digestion of municipal solid waste, journal of environmental engineering, vol. 109 (5).

T. C. Thorstensen, 1976, Practical Leather Technology 2nd edition Reinhold Publishing Corporation – New York/USA

Williams, P.T., 1998, Waste treatment and disposal, Chichester: John Wiley and Sons.

Young, J.C., 1991, Factor Affecting the Design and Performance of Upflow anaerobic digestion, Water Science and engineering.

Appendices

Appendix A: Percentage of effluents obtained in various processes (Calculated on total waste water in normal working method) (Pocket Book for the Leather Technologist, 1997)

Soaking, liming, rinsing	ca. 30 – 40
Deliming, bating, rinsing	ca. 15 – 20
Pickling, chrome tanning	ca. 5 –10
Vegetable-synthetic tanning	ca. 20 – 25
Neutralizing, dyeing, fatliquoring	ca. 15 – 20
Samming, pasting, cleaning	ca. 2 – 7
General waste water	ca. 2 – 5